# Laissez Parler:

## Freedom in the Electronic Media

David Kelley
Roger Donway

SOCIAL
PHILOSOPHY
& POLICY CENTER

ISBN 0935756-99-X



Typesetting by The Philosophy Documentation Center
Bowling Green State University

## CONTENTS

# I. Introduction

In America today, anyone who wishes to start a newspaper is free to do so, without seeking government permission. He can publish material on any subject, from any point of view, in virtually absolute freedom from government interference.

But if our would-be publisher seeks to communicate by electronic means, he will have to deal with the government at every step. If he chooses a broadcasting medium, he must get a radio or television license from the Federal Communications Commission. To win the license, he must assure the FCC of his technical ability, his finances, and his character; he must also promise, in elaborate detail, to use his station in "the public interest," and submit his program material to periodic review by the Commission.

Or suppose that he chooses cable television instead. Then he will find the right of entry determined by a battle royal among cable firms and their power brokers, all currying favor with local politicians in hopes of winning a local franchise. The weapons he will need are inside contacts, multimillion dollar "goodwill" promises about programming, political contributions, and worse.

As for entering the telephone industry: the rules and rights of this field are presently unknowable. Traditionally, the law has favored monopoly: no new entry at all. But the law is in flux, and new rules are expected soon—a condition that has existed for some years. Sooner or later these will or will not be set down in whole or in part by Congress, the Justice Department, the FCC, and/or the Federal courts. What they will say is anyone's guess.

Electronic communications, in short, is a regulated industry. There is an old name for the regulation of communications: the name is censorship. The current system does not differ, in principle, from the system of press licensure which the founding fathers regarded as a basic violation of free speech. Indeed, the Supreme Court has recognized that such regulation would be unconstitutional if applied to the print media.[1]

Why this double standard? The answer has a little to do with the physical peculiarities of the electronic media. It has much more to do with the different eras in which the two media arose. Newspapers, books, and pamphlets were already familiar vehicles of expression in the late 1700's, when sentiment for freedom was strong enough to produce the First Amendment. Telephone and radio had the misfortune to come of age during the Progressive era, when the intellectual tide had turned toward collectivism. Two documents from that era—the Kingsbury Commitment of 1913 and the Radio Act of 1927—determined the basic structure of controls which still exists. Both documents bear the characteristic marks of Progressivism: a philosophy of militant altruism, contemptuous of the individual and his freedom; and the practice of ill-concealed collaboration between established firms and the government to curtail freedom and competition.

The Kingsbury Commitment was an informal consent decree between the Justice Department and the American Telephone and Telegraph Company.[2] The Bell system had enjoyed an effective monopoly under Alexander Graham Bell's original telephone patents, but as the patents expired in the mid-1890's, competition sprang up, and by 1903 the independent companies outnumbered Bell in customers and exchanges. In 1907, however, J. P. Morgan defeated the Boston money interests for control of AT&T, and installed Theodore N. Vail as President. Vail and Morgan represented the business wing of Progressivism, preferring orderly markets and industrial monopoly, under government supervision, to the competition of open markets. Vail denounced competition as "*strife*, industrial warfare," and maintained that telephone service "must be 'One System,' 'universal,' 'intradependent,' 'intracommunicative,' and operated in a common interest. Such is the Bell system."[3]

Vail and Morgan undertook an aggressive campaign against the independents, with some success. But in 1913, the Justice Department threatened antitrust action, and the Kingsbury Commitment was the result. The commitment is often touted as a change of course, because AT&T agreed to give independent companies access to its long distance lines, and to seek government approval for any mergers with independents—thus giving up its two chief weapons in the drive for monopoly. (It also sold off Western Union, which it had acquired in 1909). But the actual effect was to freeze in place the status quo, under government regulation. Indeed, the agreement gave Vail the larger part of what he sought: a unified telephone system, and an industry stabilized by the government under AT&T's dominance. Thus the Kingsbury Commitment, together with state and Federal utility regulation, created the system of shared monopoly which still exists.

Radio legislation bore even more clearly the marks of Progressivism. The radio spectrum had functioned at first in the way the citizens' band still does, as a kind of commons used for private, point-to-point communications. With the birth of commercial broadcasting in the early 1920's, however, problems of interference grew more and more serious, and it became clear that the commons would have to be enclosed. Herbert Hoover, then Secretary of Commerce, began issuing licenses with strict conditions on frequencies to be used, power levels, and hours of operation. In 1926, however, Hoover lost a legal challenge to his authority, and abandoned his attempts at regulation. What followed were nine months of broadcasting anarchy, as stations ranged freely over the spectrum, looking for a frequency they could dominate by sheer power of output.[4]

Those nine months have been used ever since to justify the licensing system of regulation instituted by the Radio Act of 1927. But the goal of ending anarchy required only *some* means of enclosing the commons. As we shall see, the traditional means—a property system, with appropriation by users—was both possible and well-known to the legislators. If they chose regulation instead, the reason was not necessity but Progressivism. The House co-sponsor of the Radio Act said of it, "If enacted into law, the broadcasting privilege will not be a right of selfishness. It will rest upon an assurance of public interest to be served."[5] The Act nationalized the radio spectrum, and created the Federal Radio Commission to grant licenses and regulate the licensees, in whatever way it decided would serve the "public interest, convenience, or necessity." (The Communications Act of 1934 transferred these powers intact to the newly created Federal Communications Commission.) As in other industries, the public interest was largely defined by the established commercial interests, whom the regulators consistently favored over small, special-interest, church or college stations.[6]

At the time, all this might have seemed a minor power grab, the flickering out of freedom in a tiny channel of communications whose promise was glimpsed by few. But radio grew, and gave birth to television, which grew. Today the FCC finds itself, with some embarrassment, astride the greatest communications system in history. And its domain widens daily with the explosion of new electronic media: cable television, microwave distribution systems, satellite links for private communications and direct broadcasting to homes, computer data base links, teletext, videotext. This explosion has been caused in part by advances in electronic technology, and in part by the FCC's new interest in deregulation. The Commission in recent years has tried to slough off some of its power over the radio, television, and even telephone industries, and to give the new media as much breathing room as possible. But regulatory proceedings notoriously lag far behind events, and Congress has failed so far in its attempt to revise the Communications Act. More ominously still, the Commission's authority may soon extend to print media. The text of the daily *Wall Street Journal* and *New York Times* is already being sent by satellite circuits to printing plants around the country. Newspapers, catalogues, and other print services are beginning to use videotext, sending print contents to the home by telephone or cable lines; and former FCC Chairman Richard Wiley has asserted that the Commission's content regulations could be applied to videotext.[7] Thus, the merger of print and electronic vehicles will soon confront us with a basic choice: to accept government control over all communications, or to abolish regulation of the electronic media.

We are not the first to call for deregulation. The Chairman of the FCC itself has recently proposed the near-total abolition of his agency's controls over broadcasting.[8] But we feel that the case so far has not been sufficiently integrated. Deregulation is an issue of freedom, and three dimensions of freedom are at stake in communications. Economic freedom requires a system of property rights and unregulated markets in communications media, services, and products. Intellectual freedom requires a system of rights of expression for all parties involved in the exchange of ideas and information, both speakers and listeners *and* the entrepreneurs who bring them together. Political freedom requires that the voters have access to a range of information and opinion about political issues and candidates. The practical problem is to find a system which accommodates and protects all three dimensions simultaneously. But this is not possible without solving the philosophical problem of finding a common foundation, and a sound one, for the different types of freedom.

In this respect, the most serious failure in most arguments for deregulation is their disparate treatment of intellectual and economic freedom: the first is defended as an inalienable right of the individual, while the second is defended on the ground that a free market best serves the public interest, as measured by calculations of collective economic welfare.[9] The latter argument expresses an inherently collectivist philosophy, since it assumes that the good of society as a whole is a collective value which the individual is obliged to pursue, and to which his interests and his freedom may in principle be sacrificed. This is the very moral premise which produced the Radio Act in the first place, and which still animates the Communications Office of the United Church of Christ, the National Citizens Committee for Broadcasting, and numerous other "public interest" groups clamoring for the maintenance of controls.

The failure to challenge this premise has consequences. For one thing, it makes it impossible to integrate the dimensions of freedom. Rights of expression, for example, cannot be defined independently of property rights. One man's freedom of speech does not give him a right of access to another man's living room—nor, we shall argue, to another man's broadcasting station. But that argument cannot be sustained against those who would suspend the property rights of station owners in the name of free speech unless those property rights can be seen as expressions of the same basic personal autonomy that First Amendment rights protect. Another and more important consequence is that the issue of deregulation becomes a shouting match about the meaning of "the public interest." That term cannot be defined in any but an arbitrary manner, since society in fact is not an entity which can have interests, values, or goals. "The public interest" is an empty rubric that offers no objective guide to public policy; its function is merely to serve as a cover for those with some private vision of the good to enforce that vision on others. The failure by deregulation advocates to reject the concept, and the collectivist moral premise that goes with it, helps keep alive the conceptual framework in which those efforts flourish.

The individualist approach, by contrast, recognizes that values and interests are attributes of the individual. Each person's life is a value to him, as an end in itself, because it is *his*, not because it is a fragment of a collective life. Individuals are alike, however, in sharing the same basic capacities for pursuing their values and interests, and these shared capacities are the basis for social organization. Thus man's fundamental trait is reason, and Ayn Rand has argued eloquently that freedom is the basic social need for a rational being.

> In order to sustain its life, every living species has to follow a certain course of action required by its nature. The action required to sustain human life is primarily intellectual: everything man needs has to be discovered by his

> mind and produced by his effort. Production is the application of reason to the problem of survival....
>
> Since knowledge, thinking, and rational action are properties of the individual, since the choice to exercise his rational faculty depends on the individual, man's survival requires that those who think be free of the interference of those who don't. Since men are neither omniscient nor infallible, they must be free to cooperate or to pursue their own independent course, each according to his own rational judgment. Freedom is the fundamental requirement of man's mind.

The function of the concept of *rights* is to make this fundamental requirement clear and explicit. "A 'right' is a moral principle defining and sanctioning a man's freedom of action in a social context."[10] In this respect, the right to property has the same basis in the needs of the mind as the right to freedom of speech and press.

The concept of rights replaces the concept of public interest in our approach. This does not mean that there are no shared values in society. The system of rights itself, as embodied in a legal code and protected by government, is one such value. It is a single institution which we all have reason to defend, because it protects our own freedom of action, and because it makes possible the benefits of rational cooperation with others. Nor is it the only shared value. Any association—from the family to the labor union, corporation, or university—is based on the overlap of individual goals and the choice to pursue them cooperatively. But the system of rights is the only such value which society as a whole can be said to have. It is the only shared value which does not depend on a prior act of consent by those who participate, since it is the precondition of all participation by consent. And it is the only value which it is appropriate for society's collective agent, the government, to pursue by the means at its disposal: a monopoly on the right to use coercion. Hence the government is not justified in pursuing any collective value other that the protection of rights. Nor, in the nature of the case, could it be justified in abridging or suspending rights in the name of any other value, neither goodness, nor beauty, nor truth. Those are ends for individuals to seek, alone or with others, each by his lights.

From an economic standpoint, communications is an industry, one among many, to which these principles apply. The industry also supplies the channels that link minds together, and it plays a major role in the democratic process by which the people at large influence and monitor their government. In the three chapters that follow, we will look at the electronic media from each of these perspectives.

## II. Communications and Economic Freedom

Production is inherently a rational process, involving the identification of needs, resources, and techniques of producing; and the weighing of values, risks, and opportunities. In any advanced economy, the process is largely a cooperative one, but the rational judgments take place in individual minds; the risks are borne and efforts put forth by the individual agents. The physical inventions which made electronic communications possible were the work of individuals such as Samuel Morse and Guglielmo Marconi. Of equal importance were the economic "inventors"—Owen Young of General Electric, Harry Davis of Westinghouse, David Sarnoff of RCA—who saw commercial opportunities others had missed, and invented the structures of cooperation which allowed individuals and corporations to combine in exploiting the new technology. Countless others participated, each contributing something definite in the way of capital, effort, or insight.

Property rights offer protection to all participants by allowing them to agree at the outset on the terms and conditions of their participation. Such protection is crucial. The development of the industry required judgments integrating hard fact, scientific speculation, business experience and acumen, knowledge of local conditions, and a large dose of educated guesswork. The need to cope with such partial and fragmentary knowledge is common to all productive enterprises, even well-established ones, and in that context it is inevitable that rational men will come to different conclusions. Property rights protect the freedom of each individual to act on the conclusions he reaches. They prevent anyone from imposing risks on others unless he can persuade them that the risks are worth taking; but they allow anyone to pursue any idea, and undertake any risk, without having to persuade those whose cooperation he is willing to forego.

Thus the case for economic freedom is rooted in the nature of economic activity itself. It does not have to be defended by reference to "the public interest," nor by the role of economic freedom in protecting intellectual freedom. In regard to electronic communications, this means that the freedom of those who maintain and operate the physical media has an intrinsic value, above and beyond the rights which protect the contents communicated.

Our argument is structured as an answer to three problems and objections that have been raised against such freedom. (1) It is widely assumed that a system of property rights in the electromagnetic spectrum is not possible, so that wireless communications require government licensing, and thus government oversight of licensees. (2) Even if a system of property rights were possible, some feel that government regulation would be better able than a free market to ensure that the media serve the public interest. (For reasons already noted, we reject the standard of the public interest, but there is more to be said about the consequences of putting that standard into practice.) (3) Even if the FCC cannot carry out the assignment in (2), there still seem to be imperfections in the market, some people argue, specifically tendencies toward concentration and the problem of "natural monopoly" in the wire media, which call for regulation.

### A. *Property Rights in the Spectrum*

At a Radio Conference which Hoover convened in 1925, he declared: "The ether is a public medium."[11] The principle that the public owns the airwaves was incorporated into the Radio Act, and has been taken for granted ever since as a kind of axiom. The contrary principle of private property in the spectrum has been dismissed out of hand; indeed, as Ronald Coase observed, the very idea is unfamiliar to most people, including many in the industry.[12] Yet the spectrum is a physical resource, and it is a basic principle of American law that those who discover and first make productive use of a resource may acquire a property right in it by appropriation.

In this respect, the law incorporated the views of John Locke and William Blackstone. Locke's famous argument for appropriation was that each person has a kind of property right in his own person, and specifically in his labor; when he employs his labor on an unowned physical resource, the physical thing becomes "mixed" with his labor, and thus with his right; the thing comes to have "by this labor something annexed to it, that excludes the common right of other men."[13] Unfortunately, Locke's formulation put insufficient weight on the *purpose*, as opposed to the *act*, of labor. Building a sandcastle for amusement on an unowned stretch of beach is a way of "mixing one's labor" with the sand, but confers no lasting right to it; building a house there should confer such a right. The crucial point, as Blackstone recognized, [14] is that whenever the purpose of investing labor now is to reap a later return, a property right is necessary to protect this productive activity.

The metaphor of mixing one's labor may also suggest that some physical transformation of the resource must be involved. But a physical transformation is not sufficient to create a right, else leaving footprints in the forest would give one title to the path. Nor is it necessary. What matters is that the labor add *value* to the resource, and this can happen without any physical change. The productive use of the radio spectrum does not actually change the electromagnetic field: turn off the transmitter and the field remains as it was, ready for the next user. Nevertheless, the pioneers who discovered the technical and economic means of using the spectrum did bring about a relational change in it, creating a potential for communications and thus adding value where none had been before.

The traditional principle of appropriation, then, can be formulated as follows. The private appropriation of an unowned resource is legitimate, and should be recognized as a source of ownership, when two conditions are met: a) the use to which an individual puts a resource is productive, i.e., it involves the creation of value; and b) he cannot produce or enjoy that value without long-term control over the resource.

Point (b) is what distinguishes a commons from a system of private ownership. For two decades after Marconi's invention of wireless telegraphy in 1897, the spectrum was used mainly for brief point-to-point communications, as in ship-to-shore radio. During this period, the spectrum functioned reasonably well as a commons. There *were* local exceptions. Thomas Clark of Detroit, for example, began a wireless service for Great Lakes steamers in the first decade of the century, offering music and news in addition to the usual telephone links, but was put out of business by a rival using the same frequency; the law of property should have protected Clark.[15] By and large, however, problems of congestion and interference were minimal; no one needed, or could have used productively, an exclusive control over a frequency.

But starting in the 1920's, commercial broadcasters were on the air for continuous periods during the day, and in order to develop an audience and invest in program material, they needed advance certainty of being able to use the same frequency during that period every day. At this point, then, the right to appropriate frequencies, in specific geographical areas, should have been recognized by law.

(Most early stations were on the air for only a few hours each day. Since the spectrum, unlike land, is not actually changed by use, it would not harm the value produced by one station for another to use the same frequency during other hours. It would have been reasonable, then, for the law to recognize rights to certain frequencies, in certain areas, *at certain hours*—the ones during which a station had established regular service. A station wishing to expand its program would then have to contract or merge with others on its frequency.)

Had this Lockean principle been adopted by Congress, as a kind of radio Homestead Act, and applied by the courts, appropriation would be an ongoing process as technological and economic invention opened up new regions of the spectrum; today the frontier would be in the far microwave (though in sparsely settled areas, frequencies in the radio and television bands are still available). But the history of broadcasting suggests that the process would wax and wane: there would have been temporal "windows" stretching from the earliest use of a region to the development of scarcity. One such window existed in the AM radio band from 1920, when Westinghouse went on the air with KDKA, to about 1925, when Hoover stopped issuing licenses because of crowding. Another window existed from around 1935 to 1950, when television and FM radio battled for the use of the VHF band (an issue which the FCC settled by political compromise, with unfortunate results as we shall see). The brevity of these periods made the issue of scarcity correspondingly urgent.

Indeed, the Supreme Court has used the fact of scarcity as the major justification for licensing and regulation. In the landmark *NBC* case of 1943, the Court argued:

> The plight into which radio fell prior to 1927 was attributable to certain basic facts about radio as a means of communication—its facilities are limited; they are not available to all who may wish to use them....
>
> ...The facilities of radio are not large enough to accommodate all who wish to use them. Methods must be devised for choosing from among the many who apply.[16]

Yet scarcity is not in fact peculiar to radio; it is common to *any* resource traded in a market. The supplies of land and oil, of capital and labor, are also "not large enough to accommodate all who wish to use them." They are only large enough to accommodate all who are willing to pay a market-clearing price. Thus if the Court applied this argument consistently, it would deny property rights in every resource. But two factors may explain, though they do not justify, its selective application of the argument to the broadcast spectrum.

One is that the Court focused on an alleged contrast between the wireless and print media: "Unlike other modes of expression, radio inherently is not available to all. That is its unique characteristic, and that is why, unlike other modes of expression, it is subject to government regulation."[17] The appeal of this argument is that there is no obvious physical limit on the number of newspapers that can exist in an area, whereas the spectrum, like land, is a fixed resource. In economic terms, however, the argument involves a curious piece of circular reasoning. Anyone who wishes to start a newspaper must buy the resources he needs—labor, capital, and materials—by bidding them away from other potential uses. That is the only sense in which the print media are "available to all." But the only reason why someone wishing to start a broadcast station cannot similarly bid a frequency away from other uses is that FCC regulations prohibit it. The regulatory apparatus which the Court attempted to justify is responsible for the artificial asymmetry on which its argument turns.[18]

A second premise that may be lurking in the argument from scarcity is Locke's proviso that appropriation is legitimate only when "enough and as good" is left for others. Scarcity develops in a resource only when "enough and as good" is *not* left for others; it develops at the closing of a frontier, and in radio this happened with a speed that made the closure highly visible. But the proviso has no application here; its intent is only to prevent the exclusion of anyone from resources, like land or water, that are essential to his survival. The only thing from which latecomers to a given region of the spectrum are excluded is the chance to earn a stream of profits, which others had taught them how to make, without either buying a frequency in that region or finding some way to enter a new region. No one is entitled by natural right to anything more. Nor is it unfair that those with the courage and foresight to get in early, when risks were high and ingenuity at a premium, were able to obtain the resource by appropriation. Simple justice would require nothing less.[19]

But what about the content of the rights so acquired—what would it mean to have property in the spectrum? Perhaps the easiest system to imagine now, after fifty years of government licensing, and the easiest to implement as a first step, would be one that simply transforms each current license into a property right. Property would then consist in the right to generate waves within a certain band of frequencies, covering a certain geographical area. As a right, it would not be subject to periodic review, nor therefore to the uncertainty and intrusiveness of the present system. It could also be traded in a free market, so that property in the spectrum would be open to acquisition by those who value it most highly, for prices that would no longer need to include the cost of legal maneuvering in Washington.

But two further changes would allow a much greater degree of economic freedom. The first would be a change (to put it in technical language) from input to output definitions of the right. An FCC license gives a broadcaster the quasi-right to use a certain band of frequencies—the band width is 10 kilocycles in AM radio, for example, and 6000 kc (or 6 megacycles) in television. But since the bands cannot be either subdivided or combined, the licensee lacks an incentive to discover new methods (new inputs) of generating his signal. The Bell system can now fit 36,000 telephone circuits into a microwave band that carried only 2100 in the early 1950's, but broadcasting bands have not changed over that period because broadcasters, who cannot sell any part of their band, have no reason to economize on their use of it.[20] Again, if two adjacent TV channels could be combined, a station could use a 1200 line scan instead of the 525 lines now standard, allowing larger pictures with higher resolution.[21] There might well be a market for that service, but current licenses prevent anyone from finding out. Finally, a wave on one frequency will cause interference on adjacent and harmonic frequencies. There is no single best method of minimizing this problem; it depends on local conditions. An output definition of rights would allow broadcasters the freedom to reach contractual solutions based on their knowledge of those conditions.[22]

Similarly with geographical areas. FCC licenses currently specify inputs: the power each station may use, the polarization of its signal, and the height, location, and directivity of its antenna. But there are various combinations of inputs that can produce the same geographical coverage, and a right which specified the area directly would allow each broadcaster freedom to act on his judgment about the most efficient mix. As with frequency bands, it would also allow him to subdivide his area and sell off part, or to purchase adjacent areas. To be sure, the areas cannot be defined with the same precision as in real estate. An electromagnetic wave attenuates gradually, so that if it is strong enough to reach the perimeter of one area with a usable signal, it will carry over into the next, causing some interference there. A right to cover an area with a signal would therefore have to be limited by the obligation not to raise aggregate interference in adjoining areas above some threshold amount, to be settled by the law. There are technical controversies about the details of defining such rights, but there is no doubt that a considerable shift in the direction of output specifications is both possible and desirable.[23]

A second major change would be to eliminate FCC restrictions on the uses permitted in each region of the spectrum—its "zoning" plan. Zoning is supposed to simplify problems of interference, but this would not be necessary in a property system since anyone proposing to use a frequency, for whatever type of service, would be responsible for the interference he causes existing users. Another ostensible purpose is to collect similar uses together so that, for example, radio listeners will not need to purchase spectrum-wide receivers. But if that *is* the most efficient use of the spectrum—a matter that cannot be determined *a priori*—the market will provide incentives for it. Broadcasters, for example, in choosing a frequency, will hardly ignore the question whether their audience has the equipment to receive it. The real function of zoning, rather, is to substitute the Commission's judgment for that of the market as to which services are worthy of access to the spectrum. That is a straightforward abridgment of economic freedom.

It has also had uneconomic results. In the VHF band, for example, the FCC has allocated 12 channels to television. As we shall see, the stations occupying these channels are earning monopoly profits; the reason is that the zoning plan prevents the entry of any further competition. There is room for 33 additional channels in the band, and while it is impossible to foresee the outcome of a free market with any precision, it is virtually certain that some of these channels would be purchased by potential television broadcasters were they free to do so. Again, large sections of the spectrum have been reserved for government use, chiefly by the military. In the VHF band, 38% is exclusively for government, and another 14% for mixed government/private uses.[24] The government certainly has as much right as anyone else to appropriate frequencies, but a market would at least put a price on them, so that officials could make a rational decision whether to keep the frequencies, or to sell them and purchase alternative media with the proceeds.

If property rights in the spectrum are thus possible, and fully in accord with principles governing property in other resources, it is worth asking why the idea has never been adopted. Today, of course, it must compete with the well-entrenched licensing system, which has all the weight of intellectual inertia on its side. But how did licensing come to be entrenched: why was it chosen in the 1920's when *it* was the novel theory? The notion of property was certainly in the air at the time. In one court case, a station argued that it had created value in its channel, in the form of a reputation for its programming, which an interfering broadcaster was unjustly taking from it; the judge agreed, finding a precedent in trademark law. This decision was reprinted in the *Congressional Record* as part of the proceedings on the Radio Act.[25] At the same time, Hoover was frequently urged by people in the industry to permit property rights in radio channels, and the issue was being discussed in professional journals.[26] Indeed, Clarence Dill, Senate sponsor of the Radio Act, wrote afterward that Congress was conscious of overturning legal precedents in refusing to recognize property rights in the spectrum.[27] The licensing system, then, was not adopted for lack of an alternative; it reflected a conscious decision to maintain government control of the new medium "in the public interest." Both the motive and the effect of the Radio Act were to install government controls at the ground floor of the new industry before a structure of private rights could develop. It is time to consider the results of this decision.

## B. *The FCC and the Public Interest*

The Communications Act directs the FCC to act on behalf of "the public interest, convenience, or necessity." In order to do so, the Commission must: a) provide that vague goal with some definite content and find a way to make the public interest, so defined, actually effective in its deliberations; and b) find ways of gathering and integrating all information relevant to deciding which regulatory actions will best achieve that goal. But there are general grounds, amply confirmed by the Commission's actual practice, for doubting that it can succeed at either task.

The problem of information is the more obvious and widely appreciated. Ronald Coase quotes a former FCC Chairman to the effect that the Commission cannot regulate intelligently "if the industry knows more than the government does." But as Coase points out, the industry *inevitably* knows more than the government does. The FCC "cannot, by the nature of things, be in possession of all the relevant information possessed by the managers of every business which uses or might use radio frequencies, to say nothing of the preferences of consumers for the various goods and services in the production of which radio frequencies could be used."[28] Yet all this information, which *is* possessed by separate individuals, is automatically integrated by the price mechanism of a free market, whose operation the Commission is expected to surpass.

The result is that, as a rule, regulations not only fail of their intended effect, but often have the opposite one. For example, the prime time access rule, which took effect in 1971, prohibits TV stations in the top 50 markets from presenting more than three hours of network programming during prime time, 7:00 to 11:00 p.m. The intent was to give non-network producers a chance to enter this lucrative market. But there is a reason why most stations use network material: television programs are expensive to produce, and so must have a national audience in order to generate sufficient advertising revenues. In complying with the rule, stations have adjusted their behavior to that fact, not to the Commission's intent. They have filled the required non-network time with reruns purchased from the syndicated market, or with news and talk shows, which are relatively cheap to produce; they have not purchased new material from independent producers. Indeed, since the latter can sell their product more easily to networks than to local stations directly, it is likely that the rule has worked against them.

In regard to point (a) above, we have already argued that "the public interest" is a term without objective meaning, so that any interpretation of it is necessarily arbitrary. This is certainly true of the FCC's interpretation, which, at the deepest level, consists in three assumptions. The doctrine of *localism* reflects a vision of the broadcast station as a kind of small-town newspaper on the air; the FCC has consistently announced its preference for local over regional stations, individual over group owners, and local over network programming. It has also pronounced itself in favor of *diversity*, in type of programming and in point of view—and not only as between stations, but in each station's own offerings. The third assumption is a negative one: that the market is not an adequate measure of public need; in particular, that free competition for an audience will not guarantee as much news, public affairs, or cultural programming as the Commission feels the public needs. That these stipulations are arbitrary is evident from the fact that an equally compelling case could be made for the opposite theses.

But the problem runs much deeper. The formulation of such goals does not guarantee that the Commission will actually pursue them in practice. The FCC is a political entity subject to a structure of political incentives in just the way that a firm in the market is subject to economic incentives. Individual Commissioners and staff members may act on the basis of their own views about what is proper—that is a major source of the uncertainty which communications firms face in dealing with the Commission. But the Commission as a whole, over the long run, can be expected to act so as not to alienate the powers that can adversely affect it. Three such powers are especially important: the Congressional subcommittees which must recommend appropriations for it, and can suggest legislation affecting the FCC or its decisions; the groups that have standing in Commission hearings, particularly the large communications firms, and more recently, "public interest" lobbying organizations; and the U. S. Court of Appeals for the District of Columbia, which hears appeals of agency decisions.[29] The net effect of the forces these groups exert bears no necessary relation to the Commission's official goals.

To summarize: The FCC is incapable of assembling information with the same efficiency as the market, and is not likely to act on any information which does not come laden with the political incentives to which it responds. It would be an Augean task to examine each of the myriad regulatory actions which bears out the truth of that assertion, but it is worth looking at a few of the major actions which have shaped the current communications industry.

*Commercial television*

In economic terms, there are two very striking facts about the television industry. One is that there are just three networks with which most local TV stations are affiliated. The other is that profit margins are extremely high compared with other industries. Earnings for the industry as a whole (before taxes) were 21.5% of revenues in 1979; for the local stations owned by or affiliated with the networks, average profit margins were 33%.[30] Both of these facts were results, more or less unintended, of FCC actions.

When the Commission announced its final allocation plan for TV in 1952, twelve VHF channels had been granted to television, with UHF channels 14-83 set aside for future expansion. This awkward system was the product of two political compromises. On the one hand, CBS was developing a color system that worked better in the UHF frequencies, and the FCC liked that region better because it allowed more channels and so more diversity; but NBC preferred the VHF band because technology had already been developed for that region. The resulting two-tier system was to be a thorn in the agency's side for two decades. Having given TV just 12 channels in the VHF, it had to pin all its hopes for diversity on the UHF band, but those higher frequencies require about 50 times as much power to produce an equivalent signal.[31] This and other problems made UHF stations uncompetitive and they were not generally profitable until the 1970's. On the other hand, the Commission could not allocate more VHF channels to television because—this was the second compromise—FM radio sought the same region. Indeed, FM technology had been developed first and had already been stalled for a decade by FCC inaction. In the final allocation plan, FM radio was given room between TV channels 6 and 7.[32]

The resulting barriers to entry in television were further compounded by the Commission's policy of localism. For technical and economic reasons, there is a trade-off between the number of TV channels a community can receive and the number of local TV stations it can have. The allocation plan reserved numerous channels for small communities instead of allowing larger stations in regional centers to broadcast over a larger area. The result, even today, is that most people in the country can receive three or fewer VHF channels, instead of the six or seven that a regional system would allow. This means that a fourth national network is virtually impossible. Each of the three existing networks can reach at least 95% of the television homes in the country, but markets in which a fourth VHF station exists include only 33% of the national audience.[33]

The 1952 allocation plan was made without benefit of the information a free market might have supplied about the most efficient use of the spectrum. As a result, the pursuit of localism and diversity has led to a highly concentrated national system, dominated by the three networks. At the same time, moreover, the extraordinary profits which the barriers to entry have permitted commercial broadcasters have given them a vested interest in keeping the barriers in place. Their influence on the Commission has worked against even the nominal pursuit of its official policies—as the history of subscription TV attests.

At recent advertising rates, each family's attention for the duration of an hour-long program is worth only about 6½ cents to advertisers.[34] That is the economic foundation of the network system: programs of any quality require an audience much larger than any single station can supply. But many people would be willing to pay much more for programs they value, making much smaller audiences economically feasible. Thus subscription television (STV), in which the viewer pays for a descrambling device, is potentially a strong competitor with the segment of the industry supported by advertising. But the latter sought, and won, political protection. By 1959, pressure exerted through Congress by regular TV interests, and by consumer groups worried about the loss of "free" TV, led the FCC to limit the number of experimental STV stations it would allow; only a few were ever authorized.

In 1968, the Commission relaxed its prohibition, but attached severe conditions to any STV service: no series programs or commercials were allowed; no films between 2 and 10 years old, unless they had already been rejected by regular TV stations; and no sports events of a type that had been shown on regular TV during the preceding two years.[35] These were called the "anti-siphoning" rules, because their ostensible purpose was to keep STV from bidding these popular programs away from free TV, despite the fact that such programs can be duplicated: most people who would pay to watch a specific film or sports event would accept a close substitute if it were offered free.[36] The only real effects of STV would have been to draw some people away from free television, lowering advertising revenues; and to bid up the price of programming somewhat, raising costs for regular stations and networks. The threat of this (probably mild) squeeze on their profits kept the anti-siphoning rules in place for a decade. (The rules were extended to cable television in 1970. In 1977 the DC Appeals Court found against the extension, and the FCC subsequently repealed the rules altogether. By 1979, there were six STV stations operating with applications pending for 20 more,[37] but cable TV will probably absorb much of the demand there once might have been.)

### *Cable*

AT&T has long used coaxial cables to connect networks with their affiliates, and throughout the 1950's they were used in rural communities as well to link homes with a single tall antenna that could pull in distant signals. By 1962, however, cable was also being used to bring *additional* channels into communities that already had a local station, and the FCC began indirect regulation of cable under pressure from broadcasters to limit the competition. In 1966, it forebade cable systems to import signals into the top 100 markets unless they could show that doing so would not hurt the local UHF stations, which the Commission was still, these many years after its original mistake, trying to nurture. In 1968, it banned importation into these markets altogether, and the next four years saw a freeze on cable television.

The freeze and its aftermath illustrate a common pattern in government regulation of industry. A regulatory agency seeks compromise at all costs among the groups bringing pressure on it, lest any of them push to have the agency's decision overturned by Congress or the courts. When innovations threaten established firms, or threaten the power which "public interest" groups have achieved, such compromise often takes the form of delay: nothing will happen until the innovators get organized as a pressure group, and then compromise will consist in a further delay of the innovation, and the attachment to it of conditions which protect the established interests.[38] Thus during the freeze, cable firms organized politically, with some aid from the Nixon Administration's growing antagonism toward the networks, and began negotiating with broadcasters, public interest groups, and the FCC. The result, in 1972, was a comprehensive set of regulations for cable.

Cable systems had to carry any local broadcast station that desired it, and were limited in the number of distant stations they could import. Each system had to carry one channel of original material—basically a punitive measure, since original programming was not profitable until satellite links later in the decade made cable networks possible. Each system had to carry three channels of free local access, one each for local governments, educational institutions, and the general public (i.e., the media activist groups). All other channels were to be offered on a common carrier basis and the anti-siphoning rules were applied as well. The motivation for these Draconian measures was of course to protect existing broadcasting interests. Curiously, however, that was the explicit rationale as well. Under the Carroll doctrine, which the Supreme Court applied to cable in a 1972 decision, the FCC was permitted to act to protect the profits of broadcasters if it felt that the stations would not otherwise be able to afford the public service programming required by the Commission.[39] In any event, later Court decisions, and policy changes at the Commission, have repealed most of these regulations. Hence the net effect of FCC action was to delay for ten to fifteen years the explosion now taking place in cable. Owen, Beebe, and Manning estimated in 1974 that the cost of the regulations to consumers was on the order of a billion dollars.[40]

*Domestic satellites*

The success of NASA's experimental communications satellite, Echo I, led Congress to create the Communications Satellite Corporation in 1962. Comsat's initial mandate was to put a satellite system in place for international communications, and this it did within three years. But satellite links promised even greater benefits in the domestic field: they could help open up the long-distance telephone market to competition, allow cheaper means of data transmission for business, and provide the television networks with an alternative to AT&T's Long Lines for connecting with their affiliates. But the promise was delayed for ten years while the FCC sifted its way through a tangle of competing pressures that has been described in detail by Michael Kinsley.[41]

The Commission held hearings in August of 1966, but decided to put off any action until a task force in the Johnson Administration could study the subject and file a report. The impulse to spread responsibility for any policy decision was understandable, since no easy means of compromise suggested itself. It was not clear, for example, whether the law gave Comsat a monopoly on all satellite communications, or limited it to the international field. AT&T used its part ownership of Comsat to discourage it from beginning domestic operations in competition with the Bell system—at the same time that AT&T used its influence with the Commission to defend Comsat's right to a monopoly. The networks favored a system that would compete with AT&T's Long Lines, but opposed a widely-publicized Ford Foundation plan to subsidize public television with the profits from such a system. By the time that Johnson's committee filed its report, President Nixon had been elected and the Commission sat back to wait for a new committee to study the subject. Four years elapsed in this fashion.

The Commission then resumed its deliberations and four more years passed, during which: numerous firms and consortiums filed proposals for competing systems and then were invited to comment on each others' plans; the FCC's Common Carrier Bureau proposed a single monopoly system with a limited role for Comsat, which led Comsat to argue that a competitive system would better serve the public interest; the FCC told AT&T that it would be anticompetitive for the company to put up its own satellites, then that it *should* put them up because leasing from Comsat would be anticompetitive, and finally that a leasing arrangement was acceptable. It was not until July 1974 that Western Union had the first domestic satellite in operation, charging rates which, as many had expected from the outset, were considerably cheaper than the land-based alternatives.

The pattern of regulatory behavior we have described in the preceding pages has often been taken to mean that the FCC has failed to pursue the public interest. A more honest, and more accurate, interpretation would be that that is precisely what the public interest amounts to. The Commission has, after all, invariably wielded the coercive powers assigned to it by law in accordance with the play of political pressures. What else could it mean, in reality, to pursue the public interest? The problem lies not in the agency's exercise of its mandate, but in the mandate itself. It should be withdrawn.

### C. *Competition and Monopoly*

Even if we abandon the standard of the public interest, and the massive intrusions into economic freedom committed in its name, many would argue that some oversight is still necessary to deal with tendencies toward concentration which the market itself cannot control. Two types of regulation are said to be necessary, corresponding to the different "market imperfections" ascribed to wire and wireless media.

Because the broadcast channels in any area are limited, the FCC enforces certain ownership rules. No person or firm may own more than one station of a given type (AM, FM, TV) in a single market, and no one may own more than seven of each type altogether, with no more than five of the TV stations being VHF. Thus each of the networks owns five VHF stations, located in the largest markets; there are numerous other group owners such as Metromedia, Group W, etc. Because the FCC must approve station sales, it may also prevent any mergers among communications firms which it considers anticompetitive.

After an exhaustive review of the subject in its network inquiry, however, the FCC's own staff concluded that the ownership rules were "arbitrary and capricious,"[42] and often counterproductive. The limitation to five VHF stations, for example, means that no group of stations under common ownership can afford to originate its own program material; the rule thus perpetuates dependence on the networks. The "one to a market" rule makes no more sense in radio, given the number of channels available, than it would in the restaurant business. Even in television, an extra station, offering specialized programs to smaller audiences, might be economically possible in some markets if it could share facilities with an existing station under common ownership.[43] The only anticompetitive situations which the staff report considered remotely possible—one network merging with another, one firm acquiring all the TV stations in a market, or a regulated firm such as AT&T expanding vertically into an unregulated market—are cases in which the danger to competition comes from economic power created by government action in the first place.[44] The fact is that while there are obviously economies of scale in broadcasting, they are quite modest compared with other industries, and would not tend, in the absence of barriers to entry and other controls, to breed firms of unusual size or market share.

Economies of scale are much greater, however, in the *wire* media, so much so that it is commonly assumed that these services are "natural monopolies." Once the homes and the businesses in a city have been wired up, what potential competitor would risk the enormous investment of stringing up a parallel network of TV cables or telephone lines? Therefore, runs the argument, there inevitably will be just a single cable or telephone company in each area, which the government must regulate to keep it from abusing its monopoly position.

City governments have replaced the FCC as the major regulator of the cable TV industry. The natural monopoly argument would have a little more plausibility, here, if city officials were not quite so eager to create *artificial* monopolies, by awarding exclusive franchises, and to cut themselves in on the action. As the price of a franchise, cities normally exact franchise fees of up to 5% of revenues, free access channels for politicians and local interest groups with political influence, and promises to wire uneconomical parts of the city. In return, the companies are given monopoly power to charge ordinary viewers rates that will pay for these subsidies and return a handsome profit besides.[45] The competition to win these franchises has filled the papers with stories of influence peddling and municipal corruption reminiscent of turn-of-the-century streetcar monopolies. And none of it in fact is necessary. In some twenty-five cities there are two or more competing firms, often laying cables on the same street.[46] Cable TV must also compete with local broadcasters. Its movie channels compete with local theaters, its news channels with papers and magazines. Microwave distribution systems can now deliver the same services to the home without having to lay wires. Apartment complexes have put up satellite receiving dishes to bypass the cable company, and direct satellite broadcasting to homes will soon be possible. Directly or indirectly, cable TV is constrained by competition on every side.

Monopoly in the telephone industry is a more settled and stately affair. There are three basic markets in the industry: equipment, long distance, and local (intracity) service. When Theodore Vail made his arrangement with the government, AT&T had the only long distance lines; the company agreed to function as a common carrier, providing nationwide service under FCC regulation that effectively excluded competition. The Bell system did not have equivalent control over local service, and there are still, today, some 1800 independent operating companies. But state utility regulation gives each company a monopoly within its service area, with the Bell companies having much the largest markets. AT&T had a derivative hold, finally, on the equipment market, since it allowed only Western Electric phones to be attached to its lines.

This patchwork monopoly lasted until quite recently; the only major change was a 1956 consent decree in which AT&T agreed not to enter unregulated markets such as computers. But the system began unravelling in 1968. That was the year of the FCC's Carterfone decision which held that AT&T must allow its customers to use equipment produced by other manufacturers.[47] The equipment market, especially for business products such as office switchboards, is now fiercely competitive. At about the same time, MCI Corporation opened a competitive long distance circuit between Chicago and St. Louis, demonstrating that with microwave technology there is no more reason for monopoly in the intercity telephone market than in the intercity airline market. But there were two obstacles: AT&T refused its competitors access to the local phone nets;[48] and the FCC was reluctant to grant MCI the certificate of "public convenience or necessity" required for building a more extensive system. The courts have prodded the Commission to change that policy,[49] however, and an antitrust suit set off by AT&T's refusal to interconnect has recently ended in an historical consent decree. If approved by the court, AT&T will be free to compete in equipment, long distance, and any other unregulated markets it chooses to enter, but in exchange it will have to give up its local operating subsidiaries, which will continue to provide local service as regulated monopolies.

Thus the case for natural monopoly has contracted to the intracity market. It is still not compelling. Even if the local companies have the market to themselves, they face constraints on their ability to charge monopoly prices (for that, not size or market share *per se*, is the real issue). Demand is flexible, for one thing: most people consider basic service a necessity but their use of the phone beyond that is sensitive to price. Large users such as corporations also have some bargaining power, even when they cannot take their business elsewhere.[50] But in fact there is no reason to believe that, in an unregulated environment, the local companies *would* have the market to themselves. Various services which involve microwave links, and bypass telephone lines altogether, have been authorized by the FCC, and the importance of wire connections will presumably diminish in the intracity field as it has in long distance. And in 30% of the nation's households there is already a second set of *wires*—cable TV lines—which could provide the basis for competitive telephone service. In the circumstances, the danger of complete deregulation is minimal, certainly less than the twin dangers posed by regulation: that it will protect the phone companies against the new technologies, keeping the latter from developing their potential; or that it will place burdens on the local companies which keep them from competing effectively against the new technologies (just as the ICC ran the railroads into the ground because its regulations kept them from competing with trucks and airplanes).[51]

These dangers illustrate a wider point. Regulation, and the threat of regulation, have a chilling effect on enterprise, directly analogous to the chilling effect of restrictions on the press. A regulated industry may function and even grow within fixed channels, just as teaching and research may continue under censorship, so long as they remain within approved lines of thought. A regulated industry may even spawn new technology, as the achievements of Bell Labs attest. But the revolution sweeping through telecommunications is only incidentally a technological revolution; for all the refinements, the basic technology—coaxial cable, microwave transmission, even satellites— is not new. What is new is the use of this technology in *economically* innovative ways: the birth of new industries, services, and structures of cooperation and competition that cut across old divisions between markets and uproot old ways of doing business. It is this aspect of development that regulation crushes. Economic innovation is as risky, disruptive, and vital as any creative process, and as little able to flourish under government edicts. That is why we have undertaken to defend property rights and economic freedom on their own terms, valuable in themselves in the same way, and for the same reason, that intellectual freedom is valuable. But of course, intellectual freedom is also at stake in communications, as we shall see in the next two chapters.

# III. Communications and Intellectual Freedom

In the old days, censorship was straightforward. H. V. Kaltenborn criticized Secretary of State Charles Evans Hughes in a 1924 broadcast over AT&T's New York station; Hughes called AT&T executives to complain; Kaltenborn was taken off the air.[52] Things are more subtle today. The Radio Act and the Communications Act prohibited government censorship of broadcasting. In a pair of early cases, however, the Radio Commission was upheld in revoking station licenses for airing material the Commission did not approve of.[53] And in the *NBC* case of 1943, the Supreme Court affirmed the FCC's obligation to regulate content:

> ...we are asked to regard the Commission as a kind of traffic officer, policing the wavelengths to prevent stations from interfering with each other. But the Act does not restrict the Commission merely to supervision of the traffic. It puts upon the Commission the burden of determining the composition of that traffic.[54]

The FCC discharges this burden through various content regulations. In the circumstances, the use of such regulations to silence critics was bound to recur, and it has.[55] By and large, however, intimidation has been institutionalized, wrapped in the cloak of "the public interest," and defended as a necessary means of *preserving* First Amendment freedom. This last sleight of hand has been a *coup* for the censors, and to understand the logic of it we need to unravel the strands of First Amendment theory.

In constitutional law, the First Amendment is said to protect both individuals and their speech: both the individual's right to intellectual autonomy and the flow of information as a public good.[56] We will take up the second perspective in the next chapter; for now we are concerned only with the rights of the individual. What intellectual rights, then, do we possess?

*Intellectual Rights*

The most fundamental right is the one that was earliest fought for under the name "freedom of conscience." It is the right to maintain any belief without coercive sanction by government. Because it protects belief, as opposed to the expression thereof, it is a *one-party* right, held by individuals apart from interaction with others. It prohibits government from punishing a person for his belief and, on the other side, from compelling him to support any doctrine. And the latter prohibition applies to material as well as verbal support. As Jefferson put it in the Virginia Statute of Religious Liberty, "to compel a man to furnish contributions of money for the propagation of opinions which he disbelieves and abhors is sinful and tyrannical." That is why the First Amendment bans the establishment of religion, but the same principle would prohibit the government from "establishing" *any* doctrine through forced contributions.

A right of expression, by contrast, involves two parties. It could therefore be seen as a single right possessed jointly by speaker and hearer, a right to freedom from interference by third parties. Or we can think instead of two rights, one each for the speaker and the hearer. In that case, however, each right is contingent on the other person's choice to communicate. X's right as a speaker (or hearer) does not mean he can force Y to listen (or speak) to him, any more than his right to marry freely means the lady must consent. But what happens when we add a third party: the owner of a medium through which speaker and hearer communicate? If the physical medium is all he provides, there is no change in the structure of *intellectual* rights; speaker and hearer are free to communicate, contingent only on the owner's exercise of his property right. In mass communications, however, owners play a more active role. In addition to maintaining the physical plant, a publisher or broadcaster provides an intellectual service to his audience by choosing speakers they wish to hear, and to speakers by attracting an audience for them to address. Because these services involve judgments about *content*, the owner's intellectual rights are at stake. In relation to his audience, he exercises the right of a speaker; in relation to speakers wishing to use his facilities, he exercises the right of a hearer.

Now all of these are what political philosophers would call *negative* rights, because they protect freedom of action, and impose on others only a negative obligation—the obligation not to interfere. But some have held that there are also positive rights to certain goods (a job, an education, medical care), rights which impose *positive* obligations on others to provide them with the goods. In the case at hand, the positive right of a speaker would require that someone provide him with a medium and/or an audience; the positive right of an audience, on the other hand, would impose on someone else the duty to provide them with intellectual or artistic goods. The essence of FCC policy in this area is that it endorses both types of positive right.

### *The Equal Opportunity Rule and the Fairness Doctrine*

In every major statement about its content rules, the Commission has invoked "the right of the public to be informed" as the basis of policy.[57] The Supreme Court has concurred: "It is the right of the public to receive suitable access to social, political, esthetic, moral, and other ideas and experiences which is crucial here."[58] To enforce this positive right, the FCC imposes several rather nebulous obligations on broadcasters. In order to retain its license, a station must keep the Commission informed about the amount of time it devotes each week to preferred categories of programming: news, public affairs, arts, local issues, etc. The Commission does not set any precise minimums in these categories, but it does hold broadcasters responsible at renewal time for any promises they have made in their original application. The licensee is also required to engage in a process of "ascertaining" community needs by interviewing leaders in the various walks of civic life. Here again there are no precise rules, but the Commission has announced that to be on the safe side a broadcaster will want to interview leaders in nineteen categories, from government to "Recreation," and conduct between 60 and 220 such interviews (depending on the size of his city) during the license term.[59] He must also make publicly available a file showing what programs he has developed in response to community needs thus ascertained. These actions are intended to fulfill the community's positive rights as an audience.

But the community also has a positive right of access, as speaker, to the licensee's facilities; in this respect his station is "a sort of mouthpiece on the air for the community it serves."[60] This right of access is given more concrete form by two other content regulations: the equal opportunity rule and the fairness doctrine. The first was imposed by section 315 of the Communications Act; it says that if a station gives one candidate for political office time on the air, it must "afford equal opportunities" to all other candidates. The fairness doctrine, which evolved from Commission rule-making, requires that if a broadcaster deals with "controversial issues of public importance" in any aspect of his programming, he must give "reasonable" coverage to the opposing viewpoints on the issue.[61] The fairness doctrine is thus a far more sweeping stipulation. During the 1960's, however, the FCC developed two narrower rights of access under the fairness doctrine. The editorializing rule says that if a station endorses a candidate in an election, his opponents must be notified and given time to reply; if it opposes the candidate, he himself must be given that opportunity. The personal attack rule extends the latter right of access to any individual or group whose "honesty, character, integrity or like personal qualities" are attacked on the air.[62] It is important to note, as Benno Schmidt points out, that the purpose of the personal attack rule is not protection of the individual attacked. That is the function of libel and defamation laws. The FCC rule applies only in the context of controversial issues, and the rationale is that "when an attack occurs during discussion of such an issue, the target of that attack is the best person to give the public another point of view."[63]

These regulations require large scale intrusions into broadcasters' freedom to act on their judgment in choosing content. To that extent they are obviously incompatible with broadcasters' rights of expression, and both the courts and the Commission have pretty well admitted as much.[64] Their defense has been that the regulations increase net freedom of expression for society as a whole. Like most claims about that nebulous beneficiary, however, this one melts away under analysis. The actual effect of FCC controls has been to chill and distort the marketplace of ideas no less than the economic market. And, as we shall see in concluding this chapter, that is precisely what a closer philosophical look at the positive rights rationale would have led one to expect. As with money, bad rights drive out good.

One effect of the rules is to give private pressure groups considerable power over broadcasters. Until recently, licenses in radio and television ran for three years,[65] but were usually renewed automatically, without a hearing, unless the Commission had evidence of serious lack of compliance with its public service requirements. Under Commission rules, however, a renewal hearing must be held if another applicant seeks the station's license; a hearing can also be triggered if a group not seeking the license for itself files a complaint that the station has not operated in the public interest. Both mechanisms have been used to force stations to alter their programming.

In 1968, the owner of two radio stations in Atlanta applied for permission to sell the stations to a new owner who planned a format change from classical to more popular music. The FCC turned down a protest petition from a listeners' group which opposed the change, but the Appeals Court ordered the Commission to hold hearings. Thus began a ten-year battle between the Court and the Commission over radio format suits, which was finally resolved by the Supreme Court in the Commission's favor. In the interim, however, stations like KXA-AM in Seattle were forced to compromise with special interest groups under threat of a challenge to their licenses, and thus abandon programs popular with wider audiences. Sales of stations were similarly quashed by the threat of expensive legal actions against changes the buyers wanted to make.[66]

But the really fertile ground for these pressures has been in civil rights and minority programming. In a case that dragged on through the late 1960's, WLBT, a TV station in Jackson, Mississippi, lost its license for refusing to air viewpoints opposed by its pro-segregationist owners. This was the first clear case in thirty years of a station losing its license for the moral-political character of what it broadcasts; there have been several such cases since. But WLBT was a landmark case on another count as well. The Appeals Court had ordered the FCC to give standing to the United Church of Christ, whose Office of Communications had originally petitioned against WLBT.[67] This precedent set off a wave of similar interventions by the United Church of Christ and other activist groups. Their petitions rarely end in license revocations, nor is that their intent. As Schmidt explains, "The primary tool on this front has been settlement agreements, in which challenges to renewal are withdrawn in return for licensee promises concerning employment and programming."[68]

Racists are not attractive candidates for First Amendment protection. Neither are Nazis, nor Communists, nor pornographers. But a right of expression cannot be made contingent on the truth or popularity of the doctrines expressed, without ceasing to be a right. Nor did the violation of WLBT's right bring about any counterbalancing gain in freedom. The United Church of Christ was free all along to express its opposing view in other media—it was even free to buy a competing broadcast station, though FCC barriers to entry made this more difficult. The effect of its petition, and almost certainly its motive as well, was to force WLBT to neutralize its own opinion by broadcasting the Church's along with it. More generally, as NBC producer Bill Monroe put it, the effect of FCC rules in this regard "is to institutionalize the broadcaster's vulnerability to any pressure group that can scrape up 35 members."[69]

Another chilling aspect of FCC content regulations is their discretionary nature. The rule of law requires, at a minimum, that the law be knowable in advance by those liable to its sanctions. Administrative law, by contrast, has aptly been described as "regulation by raised eyebrow." As noted, the Commission offers nothing more than flexible guidelines on the amount of time to be devoted to preferred categories of programming, and on the number and type of interviews to be conducted in ascertainment of community needs. These choices are left to the discretion of the broadcaster, but the Commission reserves the right to review, after the fact, his exercise of discretion. Similarly, "the licensee, in applying the fairness doctrine, is called upon to make reasonable judgment in good faith on the facts of each situation,"[70] but of course the Commission once again has the final say, after the fact, on whether a judgment was reasonable and in good faith. It may well be that the FCC adopts this policy in order to minimize its intrusion into broadcasters' freedom, but the actual effect is to maximize it, since the licensee, trying to guess the Commission's intent in the face of uncertainty, is always induced to err on the side of safety.

This problem of uncertainty cannot be eliminated. Content is an intangible thing, and does not permit the degree of precision possible in other areas of the law. When does a comment about someone become an attack? How does one determine whether an attack has occurred in connection with discussion of a controversial issue? How does one determine whether an issue *is* controversial, or isolate a single issue from others? How many opposing views on an issue must be aired? Must they all be aired at the same time of day? Must they reach the same audience? How much time can elapse between the original and the "balancing" programs? The Commission has no fixed answers to these and other questions. Steven Simmons found in interviews with officials that different rules of thumb are used by different people, and FCC case law on the subject is tortuous and contradictory.[71] In a 1971 case, then-Chairman Dean Burch offered this advice to broadcasters: "I fear that, under the present circumstances, both licensees and the public can only fall back on prayer to divine the Commission's intent."[72] *Freedom of expression needs stronger securities than prayer.*

Even if the Commission could define its rules with mathematical precision, however, another and equally chilling effect would remain.

Both the equal opportunity and the fairness doctrine are contingent rules. The first applies only when the licensee allows one candidate for election to use his facilities; then and only then must he allow equal use by other candidates. Section 315 of the Communications Act says explicitly that "No obligation is hereby imposed upon any licensee to allow use of its station by any such candidate."[73] The fairness doctrine requires that some time be devoted to discussion of controversial issues, but this affirmative obligation has been enforced only once.[74] In practice, the fairness doctrine too is contingent: *if* a broadcaster chooses to discuss a controversial issue, then he must air the opposing view. The major effect of these contingent rules is to discourage licensees from offering free time to major political candidates, and from discussing controversial issues, in the first place.

The equal opportunity rule was lifted during the 1960 campaign, so that Nixon and Kennedy could debate without triggering an equal time demand from the fourteen fringe candidates. In that campaign, the networks gave over 39 hours to such free appearances; in the next two elections, when the rule was in force, they gave 4½ and 3 hours respectively.[75] As for the fairness doctrine, the FCC received 223 complaints against stations in 1960. In 1970, after a decade of Commission activism under the doctrine, the number was 1736, and it grew to 3570 in 1975.[76] These complaints can lead to a letter of inquiry from the Commission to the station, and, if the reply is not satisfactory, to Commission action: a fine, renewal hearing, a revocation of the license. None of this encourages the treatment of controversial issues.

To be sure, these penalties are rarely imposed. Between 1973 and 1976 there were 13,800 complaints (leaving aside a block of 36,000 letters aimed at a single program). Steven Simmons notes that these produced only 244 letters of inquiry to stations, which in turn led to only 54 adverse rulings.[77] It was apparently such figures which led an officer of the United Church of Christ Communications Office to assert, in Congressional testimony, "It is inconceivable that the pale threat of fairness complaints should thwart a broadcast licensee in its search for truth and its exercise of free speech."[78] That comment does not reveal an exalted power of conception. Each of the 54 adverse rulings would be widely known in the industry, chilling many besides the victims. Each of the station inquiries had to be answered by calling in the producer, double checking his research, checking program logs for balancing material, and having the lawyers draft a reply. A small station spent $20,000 and 480 personnel hours defending itself (successfully) against such a complaint. NBC spent over $100,000 to fight, and win, a fairness action brought by Accuracy in Media against a single documentary.[79] These are obviously strong incentives for staying away from controversial matters.

The Supreme Court gave a strong endorsement to the fairness doctrine in the famous *Red Lion* case of 1969. Radio station WGCB, in Red Lion, Pennsylvania, was carrying the syndicated commentary of right-wing fundamentalist Billy James Hargis. In one 1964 broadcast, Hargis attacked a liberal writer, Fred Cook, who had previously attacked Hargis in an article for the *Nation*, and had written an anti-Goldwater book during the campaign. Cook was offered reply time at the standard rate, but demanded free time on principle; when the station refused, the matter went to court.[80] Lawyers for WGCB, and for an industry group which brought a parallel suit joined to *Red Lion*, argued that this application of the fairness doctrine would have a chilling effect, for reasons such as those given above. The Court dismissed these fears as "at best speculative,"[81] and found for Cook.

It is ironic, then, that facts about the case which came to light after the decision provide the most graphic and *unspeculative* illustration of the chilling effect. Fred Friendly discovered that Cook had been working with Democratic Party officials in a campaign to suppress right-wing opinion on radio by means of the fairness doctrine; Cook's *Nation* article, Goldwater book, and possibly even his suit against WGCB were products of this effort. The strategy was to demand free time, under the personal attack provision, for individuals named in right-wing broadcasts. The motive, according to one participant, "was to use the Fairness Doctrine to challenge and harass right-wing broadcasters and hope that the challenges would be so costly to them that they would be inhibited and decide it was too expensive to continue."[82]

What is shocking about this episode is not so much the political dirty tricks, as the Court's failure to grasp the potential for them which the fairness doctrine creates. The Court has consistently protected the print media against such dangers; five years after *Red Lion*, it struck down a Florida law which imposed a kind of personal attack rule on newspapers.[83] To defend its unequal treatment of the media, the Court relies on the scarcity argument used to defend FCC regulation in general, an argument we have already seen to be invalid. The real basis for content regulations, however, is the premise of positive rights. A closer look at these rights will show why their enforcement is necessarily destructive of freedom of expression.

## *Content Regulations and Freedom of Expression*

The argument for positive rights is always that mere freedom of action is not enough; such freedom is not effective, and so not valuable, unless the opportunity to act and the means of succeeding in action are provided along with it. What freedom of speech do we really have, it is asked, if the means of speaking are not available? The fallacy in this argument lies in a kind of wishful thinking. The obstacle to genuine freedom of action is coercive interference by others, and government can remove this obstacle by making and enforcing appropriate laws. But opportunities, and the goods produced by exploiting them, require thought and effort in the face of problems, risks, uncertainties. Thought and effort cannot be endowed—nor problems, risks, or uncertainties removed—by legislative action. They can only be redistributed. One may prefer a redistributionist society, that is, a *collectivist* society, over a free one. But there is a gulf between them that cannot be bridged, and should not be blurred, by inventing a new type of rights. In this respect, the concept of positive rights is an attempt to give redistribution the moral prestige of freedom.

Positive rights of expression obviously require a material redistribution: the owner of the medium is forced to use his facilities in ways he would not have chosen were it left to his own judgment, including the promulgation of opinions he may not share. This is itself a violation of his freedom of expression, of the type Jefferson denounced as "sinful and tyrannical." But the redistribution goes even deeper. Positive rights of expression are a claim on the intellectual as well as the material resources of others.

The Court has assigned to the public as listeners the right "to receive suitable access to social, political, esthetic, moral and other ideas and experiences." Who is to provide these ideas and experiences? The Court did not say. But the owners of the media are clearly identifiable sources, and the FCC places on them the intellectual burden of discovering what the public needs, through ascertainment procedures, and then exercising their intellectual and artistic judgment in finding material to satisfy these needs. This is not a state of intellectual freedom, nor even of honest suppression. It is a condition of forced intellectual performance, subject after the fact to punishment if it fails to satisfy the whims of the Commission or the interest groups it has given standing.

The right of access for speakers involves the same kind of intellectual expropriation. The problem in communications today is not how to get a message through, but how to win the attention of people who are bombarded with messages. Correspondingly, the positive right of access is a claim not merely on the technology of sending messages, but on the attention, the *minds*, of those who receive them. Jerome Barron, the chief exponent of access rights, is quite explicit about this in arguing that the rights apply with greatest force to the most influential media, "those media where the greatest public attention [is] focused."[84] This is to assert that the minds of the listening public are a resource to which speakers are entitled, as are the intellectual and artistic abilities of media owners who have attracted the attention of the public to their channel. What Martin Mayer said about community access in cable television is applicable more generally:

> [A]ccess to *media* means nothing at all....Access to *audience* might have some value....[B]ut access to audience must be earned, with talent. There is something bittersweet funny about the sight of all these groups of ardent young lawyers and graduate students and junior executives at foundations, none of whom can write a song anyone would sing or a book anyone would read or a play anyone would act, none of whom holds a position which gives his thought significance in the lives of others or could gather twenty-five people to hear him speak at a meeting—"demanding" access to the great audience of an entertainment medium.[85]

These demands for redistribution obviously require a sustained intrusion into the freedom of broadcasters. They also compromise the freedom of the speakers and listeners with whom broadcasters would otherwise have contracted, were they not subject to content regulations. And for what? Not to increase the *freedom* of the favored groups, but only to confer on them certain *goods*, intellectual as well as material, which they are unwilling or unable to earn for themselves. It is little wonder then that the regulations have had so destructive an effect on freedom of expression in the electronic media. *They are ripe for repeal.*

## IV. Communications and Political Freedom

Free speech theory has not been a province of pure individualism. First Amendment freedoms, like property rights and economic freedom, have also been defended on the collectivist ground of the public interest; indeed, this has been the dominant line of justification in 20th-century constitutional law. The collectivist argument, as Benno Schmidt describes it, is that "individual speech must be protected because it brings diversity, competition, and therefore efficiency to the collective search for truth, not because it is a manifestation of personal autonomy on which the state may not impinge."[86] Given the philosophy outlined in our introduction, this is not a justification we can accept. *Some* individuals, alone or in association, pursue various kinds of truth by various means, and individual rights of expression protect their freedom to do so. But society as a whole is not engaged in a "collective search for truth," or in the pursuit of any other collective goal besides the preservation of rights.

Nevertheless, there is one consideration of this type to which we are committed by the assumption that a democratic procedure for choosing government officers and policies is an essential political means to the end of preserving rights. A democratic system cannot function unless the public has access to a certain range of information, and diversity of opinion, about the candidates and issues they must vote on. If a free and private marketplace of ideas provides such information better than the alternatives—and there is no doubt that in general it does—then we have another reason for valuing it. But we also have another ground on which interventions in that marketplace might in principle be defended. We must consider the possibility, in other words, that the political implementation of rights requires some restriction on actions that would otherwise fall within the sphere of individual rights—as the jury system (assuming it is a necessary mechanism in criminal law) limits the freedom of citizens to spend their time as they choose.

But there are two severe constraints on any such argument. First, the range of information at issue here is quite narrow. The intellectual rights of the individual protect thought and discussion on *any* topic—in philosophy, art, science, and history as well as in current political affairs—whereas the argument from democracy applies only to the latter sphere. Secondly, information about politicians is obviously the sort of information which politicians have the greatest incentive to suppress. Hence any argument for a mechanism which would give the government power, direct or indirect, over the flow of information about current political affairs carries an even heavier burden of proof than would a parallel argument in any other sphere of communication.

The FCC's content regulations on broadcasting certainly do not pass this test. We have already seen how the Democratic Party used the contingency effect of the fairness doctrine to suppress opposition. The equal opportunity provision is also a contingent rule, and we have seen that its effect is to discourage stations from giving air time to candidates. Then why not make this an affirmative obligation, requiring the sale of at least some time to candidates? Congress did just this in the Federal Election Campaign Act of 1971: one provision of the bill amended the Communications Act (sec. 312 (a)(7)) to allow license revocation if a broadcaster willfully or repeatedly refuses to sell political advertising time during a campaign for Federal office. But there is no reason to believe this will increase the ability of candidates to make their views known to the public, since broadcasters in an unregulated market would hardly be reluctant to sell time for political ads. And it creates a serious danger of political intrusion into content decisions. There is the danger that the FCC's exercise of discretion under the law (it must decide, for example, whether a station is reasonable in turning down a given request for time) will be biased by the Commissioners' attitudes toward the candidates in question, especially toward an incumbent President or Congressman who has political power over the Commission. And there is the danger that candidates will wield such an access right selectively against broadcasters who oppose them, in a deliberate attempt to chill the opposition.

The Supreme Court ruled on the amendment in 1981.[87] When the Carter/Mondale campaign committee tried to buy half an hour of prime time in December of 1979, all three networks refused. At that stage of the campaign, with the election almost a year away and the first primary several months off, there were still fourteen other candidates in the running for major party nomination, and the networks feared that equal time demands would cause massive disruption in programming. CBS offered several five-minute spots, however, and the others promised time later in the campaign. The FCC held that the networks were unreasonable not to give President Carter the time he wanted, and the Court found that the Commission had acted within its authority. Doubtless it had—and that is our objection to the authority given it under the law. For what was at stake here was not the viability of democracy, even in a particular race, but only the political interest of a sitting President facing an uphill battle for reelection. It is not surprising that the Commission's vote on the action split along party lines, with the four Democratic Commissioners favoring the action and the three Republicans opposed. The case seems tailor-made to illustrate Justice Stevens' point, in dissent, that the policy "creates an impermissible risk that the Commission's evaluation of a given refusal by a licensee will be biased—or will appear to be biased—by the character of the office held by the candidate making the request."[88]

*Concentration and Public Opinion*

There is a much broader argument, however, with adherents across the political spectrum, to the effect that tendencies toward concentration in the private marketplace give a few individuals or corporations an undemocratic measure of control over the flow of information, and that, therefore, some regulation is necessary to counteract these tendencies. These "Citizen Kane" considerations will occupy the remainder of this chapter.

At the end of 1980, 7810 commercial radio stations were on the air, along with 517 VHF and 230 UHF commercial television stations.[89] There were also 1744 daily newspapers and about 7000 weeklies.[90] Even discounting for the growth of population, those figures represent a substantial increase over the 37 papers in regular publication on the eve of the American revolution.[91] Nor do those figures include books, magazines, cable and satellite channels—not to mention universities, institutes, think tanks, and all the other specialized services that make up the burgeoning knowledge industry. But media watchers have been alarmed by other indices of growth. All but about 30 VHF TV stations are affiliated with one of the national networks, thus serving as conduits for news and opinion from just three sources. The networks and the fifteen stations they own and operate regularly account for about half of industry profits. The number of cities with competing daily papers has declined from over 500 in 1923 to 34 in 1979; over that same period, the percentage of copies sold in cities with competition has declined from 90% to 30%. (The trend is certain to have continued with the recent demise of papers in Cleveland, Minneapolis, Washington, and Philadelphia.) In 1923, only 153 newspapers were owned by chains; by 1976, chains owned 1061 out of the 1765 dailies then in existence and had 70% of the circulation. At the same time, newspapers owned local broadcast stations in over 200 cities, and in 1975 the FCC identified 16 cities in which the only newspaper owned the only local station.[92]

This economic trend has been accompanied by much commentary of an alarmist nature about the power of the major media—the networks, the newsweeklies, the big city dailies—to mold public opinion. The claim is heard most often on the farther reaches of the right and left, where it is stimulated in part by a perception of the moderately liberal slant of these media.[93] But the claim has support across the political spectrum. In 1974, at the height of Watergate fever, a survey of opinion "leaders" found that TV was ranked ahead of the Presidency as the most influential institution in the country.[94] Over the last decade social scientists, looking for more quantitative evidence of power, have found some correlations, albeit indirect and partial ones, between media content and public opinion.[95]

Such correlations provide the only *objective* evidence concerning media power. But they leave open the crucial question of causality: is public opinion literally determined by the media, or do people make up their minds independently, with the correlations reflecting voluntary agreement? It seems obvious to us that both can happen. Some people are creatures of what they hear, others have independent minds. The premise of democracy is that there are enough people of the latter sort to make self-governance possible, and we share that premise. The question, then, is whether the economic trends have any effect in making independent judgment more or less difficult.

In order to answer the question, we need to distinguish two aspects of concentration: the increasing size of media organizations, and the (alleged) decrease in competition among them. The argument from size is easily the more dramatic. The largest corporations in the industry have media sales in excess of $2 billion yearly; network news programs reach audiences of 25-30 million homes. But size *per se* is of little relevance. Large revenues may allow a mass communications firm to package its views more persuasively, but in a competitive market, with firms of roughly comparable size, that power is distributed across the range of viewpoints represented, neutralizing the gain that size confers on each. The channel controlled by a large firm may also allow a single speaker to address many listeners at once, but that fact has no bearing on the speaker's influence over a *particular* listener so long as that listener has other speakers to choose from. The phenomenon of size, in other words, is irrelevant apart from the phenomenon of competition and diversity.

And the figures provided above do not in fact reveal any unqualified decrease in diversity. The existence of a few cities with single newspaper-single broadcast station links, and the declining number of cities with competing newspapers, suggest that competition has declined somewhat at the local level. (Even this is not clear. Newspapers have consolidated in the face of competition from broadcasting, as the railroads consolidated to meet competition from other vehicles. Most newspapers now compete with two or three local TV, and a dozen or more radio stations, and, on the other side, with a growing number of suburban papers.) But there has been a marked increase in competition among suppliers of national news. In 1923, when most readers had several local papers to choose from, those papers provided most of the national news available, and most of that came from the AP or UPI wire services. Today the local paper must compete with the networks, the newsweeklies, three all-news cable channels, and even a few large dailies—such as the *New York Times* or the *Wall Street Journal*—that have gone national.

In economic terms, then, the increasing size of media firms has not decreased competition; it only reflects a shift in the locus of most intense competition from the local to the national level. But the issue at hand is not so much economic as intellectual diversity—diversity in *content*—and the two do not coincide. Indeed, one central complaint against the media has been that despite fierce competition for ratings and revenues, the major media offer little variety in the type of news or range of opinion they offer. What this complaint usually means is that the views offered by the major media tend to cluster around a point slightly left of center. To that extent, the fact complained of is a natural outgrowth of our intellectual climate. The dominant media naturally reflect the dominant or centrist position on most issues, with a slight bias leftward in the direction of the center of gravity among intellectuals, with whom journalists tend to identify. Restrictions on ownership will not alter that fact; they will only distribute the bias among a larger, and economically less efficient, number of media outlets. And content regulations would be equally useless, since the regulatory agency which must apply them is bound to be staffed by people with the same general outlook as those in the major media.

*FCC Regulation and the Narrowing of Public Discourse*

Both types of regulation have done positive damage, however, creating an even tighter Establishment in television than elsewhere. In a free market, the owner of a medium can restrict the range of views he offers only at the cost of limiting his audience and, therefore, his revenues. Many individual owners choose to do just this out of the conviction that a certain point of view is right; and it is an essential function of freedom of expression to protect this choice. But as the organization grows, the pressure grows with it to earn a profitable return on invested capital, and corporate ownership (which is more sensitive to that pressure) is likely to replace personal ownership. For both reasons, size would tend to bring with it more representative coverage of diverse views as a means of attracting a larger audience. Freedom of entry into a market also makes it possible to introduce radical ideas that are not well represented by the established firms. The FCC's economic regulations, however, have diminished these market forces by creating barriers to entry, and allowing established broadcasters to earn profits far above market levels. The size of the audience which the networks have already lost to cable TV (some 15-20%) indicates the kind of demand for diversity which FCC regulations pent up.

As for the Commission's content regulations: we have already seen how they discourage the open expression of opinion. But there are two other points worth noting in this context. The first is that the fairness doctrine creates what Michael Robinson calls "the bias of artificial balance."[96] Broadcasters know that even if they cover a point of view without endorsing it, they must find at least one opposing view to balance it. But the Commission has never found a station unreasonable for presenting *only* two points of view.[97] Hence the world as structured by the doctrine is a dichotomous one. Even when there is little real debate on an issue, "the journalist is obliged to invent the second point of view,"[98] artificially conveying a sense of conflict and controversy. On the other hand, complex issues are reduced to simple oppositions of yea and nay, liberal and conservative, business and labor, etc. And this effect is inevitable, since it would impose an impossible burden on broadcasters, vastly magnifying the contingency effect, to require that they cover *every* point of view.

The second point worth noting is suggested by the FCC's behavior in enforcing the fairness doctrine. Of the 54 adverse rulings against stations which Steven Simmons counted between 1973 and 1976, 38 involved the editorializing or personal attack provision; only 16 concerned other matters.[99] That is to say: the Commission has devoted most of its attention to issues about particular persons, but has shied away from the much wider realm of abstract issues. It is not hard to see why. There is no neutral, objective way to define abstract issues, or to classify and quantify the possible positions about them. To say what the issues and alternatives are is already to have taken a stand on the matter. Even the quantitative methods of content analysis, employed by social scientists, are valid only within strictly limited contexts; and even then they often reflect, in a fairly obvious way, the researcher's own position about the issue he is studying. Hence the FCC has no way to regulate debate without joining it.

One issue much in the news, for example, has been monetary policy, especially its effects on inflation and interest rates. The issue is usually presented as a choice between easing or tightening the monetary screws, and that is indeed the debate, vastly simplified, between mainline conservative and liberal economists. But radical free market proponents, such as the Austrian economists, hold that the real issue is whether the government should control the money supply at all. They see the debate between mainline liberals and conservatives as a local dispute within a single school of thought, and they could perfectly well argue that it violates the fairness doctrine to present only the views of that school. Marxists, on the other hand, claim that the real issue is not *how much* credit to create, but who should control its allocation. Private control, they argue, will necessarily cause malinvestment and recession, and the local dispute between mainline liberals and conservatives is merely a tactic to divert public attention from the real alternatives. In order to enforce a "balanced" presentation of the issues, then, the FCC would first have to side with one or the other of these ideologies about what the real issue is: and there is little doubt that it would side with the mainline economists. It is to the Commission's credit that it avoids enforcing the fairness doctrine in such matters. But it cannot escape the fact that its power to review broadcasters' choices at renewal time creates an incentive for them to play it safe by accepting the mainline definition of the issue.

Another big issue of the past decade has been the media themselves. Most of the commentary has dealt with matters raised in this section, and most of it has been true—to some extent. There *is* a tendency toward larger size—and a flurry of activity among smaller enterprises, in electronic as well as print media. There is an unmistakably liberal slant in the major media—but those inclined otherwise can easily find news and opinion more to their taste. The media doubtless have some significant impact on public opinion (why would people read or listen if they got nothing from it?), but the views of journalists and editors are themselves influenced by trends among intellectuals and among the public at large. The marketplace of ideas, like an economic market, is an integrated system that seeks equilibrium in the face of constantly changing input, by means of various balancing mechanisms. But there is more than a parallel between the two markets. Some of the most important balancing mechanisms in the market for ideas depend on the fact that it is itself an economic market. It is little wonder then that where government intervention has had the greatest impact—in television—debate is least robust, and least diverse. *Democracy would be better served by private ownership and unrestricted markets in the media.*

# V. Conclusion

We have argued that only a free market in communications media, services, and products, based on the recognition of property rights, can accommodate and preserve the different dimensions of freedom: economic, intellectual, and political. How do we get there from here?

Some policy implications follow directly from the principles defended above. First, the fairness doctrine and the other content regulations should simply be repealed. Secondly, Congress should work with the FCC to institute a system of property rights in the spectrun along the lines indicated above: transforming current licenses into property rights, and then shifting as far as possible to output definitions, and ending the zoning rules. Finally, the other restrictions on entry, ownership, and the conduct of business should be lifted.

Two other categories of issues cannot be settled so directly. On the one hand, there are problems of eliminating conflicts among rights which cannot be solved merely by appeal to the principle of rights itself. At present, for example, there is a conflict between the right of individuals to set up satellite receiving dishes on their property, and the rights of cable and subscription companies to charge a fee for material sent by satellite. Our judgment would be that the private ownership of receiving equipment is too well-entrenched to permit a sweeping ban on satellite dishes, but that the law should forbid unauthorized trade in descrambling devices. In general, however, issues in this category will turn on considerations of historical practice, and of transaction and enforcement costs, which are too complex for treatment here.

A second set of issues we have left aside are those arising from distortions in the market caused by past regulations. The largest problem here is AT&T's dominance in telephone service. We feel that the recent consent decree, splitting the parent company from its local operating subsidiaries, is a reasonable corrective. But the divorce should not prohibit any of the parties henceforward from entering any market it chooses. More generally, while a swift return to a market system may cause disruptions, or confer advantages on established firms, that would never have occurred had a market operated from the outset; an immediately restored free market is likely to overcome such problems more quickly and efficiently than any gradual plan administered by the government. The burden of proof, at least, lies squarely on the shoulders of gradualism.

If these problems can be solved—and certainly they can—then we may finally enjoy the freedom to communicate through the electronic channels that science has laid at our feet.

# NOTES

1. See notes 16, 53, 54, and 83 below.

2. Contained in a letter from AT&T Vice-President N. C. Kingsbury to Attorney General A. G. James McReynolds.

3. AT&T Annual Report, 1910. See also John Brooks, *Telephone* (New York: Harper & Row, 1975), and Henry M. Trebing, "Common Carrier Regulation—The Silent Crisis," *Law and Contemporary Problems 34* (1969).

4. *United States v. Zenith Radio Corp., 12 F2d 614 (1926).* See also Attorney General Donovan's opinion on Hoover's authority, 35 *Ops. Att'y Gen.* 126 (1926). On the period of anarchy, see Philip T. Rosen, *The Modern Stentors* (Westport, Conn.: Greenwood Press, 1980), 101-102; and Erik Barnouw, *A Tower in Babel* (New York: Oxford University Press, 1966), 190.

5. Rep. Wallace H. White, 67 *Cong. Rec.* 5479, March 12, 1926.

6. Rosen, 103; Barnouw, 260-61.

7. Cited in John Wicklein, *Electronic Nightmare* (New York: Viking Press, 1981), 138-139.

8. Mark Fowler and Daniel Brenner, "A Marketplace Approach to Broadcast Regulation," *Texas Law Rev. 60* (1982).

9. E.g., Fowler and Brenner, *passim*; and Arthur S. DeVany *et al.*, "A Property System for Market Allocation of the Electromagnetic Spectrum," *Stanford Law Rev. 21* (1969), esp. at 1500.

10. Ayn Rand, "What is Capitalism?" in *Capitalism: The Unknown Ideal* (New York: New American Library, 1967), 17; "Man's Rights," in *Virtue of Selfishness* (New York: New American Library, 1964), 93. The reader may consult these works for a fuller statement of the position taken in the text.

11. Herbert Hoover, *Memoirs*, vol. II: *The Cabinet and the Presidency* (New York: Macmillan, 1952), 144.

12. Ronald Coase, "The Federal Communications Commission," *Journal of Law and Economics 2* (1959), 17.

13. John Locke, *Second Treatise of Civil Government*, sec. 27.

14. William Blackstone, *Commentary on the Laws of England*, II 1.

15. Barnouw, 33-34.

16. *National Broadcasting Company v. U. S.*, 319 U. S. at 213, 216 (1943).

17. Ibid., at 226.

18. See Fowler and Brenner for a more detailed critique of the scarcity argument.

19. Cf. Ayn Rand, "The Property Status of the Airwaves," in *Capitalism*.

20. Charles Lee Jackson, "The Allocation of the Radio Spectrum," *Scientific American 242*, #2 (1980), 38.

21. R. G. Noll, M. J. Peck, and J. J. McGowan, *Economic Aspects of Television Regulation* (Washington: Brookings Institution, 1973), 263.

22. Jora Minasian, "Property Rights in Radiation," *Journal of Law and Economics 18* (1975).

23. In addition to Minasian, see DeVany *et al.*; and Harvey J. Levin, *The Invisible Resource* (Baltimore: Johns Hopkins University Press, 1971), 89-104.

24. FCC, Network Inquiry Special Staff, *New Television Networks* (1980), I 45.

25. *Tribune Co. v. Oak Leaves Broadcasting Station* (Circuit Court, Cook County, Ill., 1926); reproduced at 68 *Cong. Rec.* 216 (1926).

26. Hoover, *Memoirs* II, 139-40; Coase, 30-31.

27. Clarence Dill, *Radio Law* (Washington College Press, 1938), 77-78.

28. Coase, 18. This argument, generalized to the economy as a whole, has been used by Austrian economists against the idea of government planning. See F. A. Hayek, "The Use of Knowledge in Society," *American Economic Rev. 35* (1945).

29. See Erwin G. Krasnow and Lawrence D. Longley, *The Politics of Broadcast Regulation* (New York: St. Martin's Press, 1973), ch. 2-3.

30. FCC, *Annual Report for FY 1980*, 29-30.

31. FCC, *New Television Networks*, I 72, II 70-71.

32. See Krasnow and Longley, ch. 5; and Don F. Erickson, *Armstrong's Fight for FM Broadcasting* (University, Ala.: University of Alabama Press, 1973).

33. Noll *et al.*, 100-101, 116; FCC, *New Television Networks*, I 43-50, 80.

34. Ibid., I 38.

35. Ibid., I 110-19.

36. Richard Posner, "Appropriate Scope of Regulation in Cable TV," *Bell Journal of Economics and Management Science 3* (1972).

37. FCC, *Annual Report for FY 1979*, 31.

38. Noll *et al.*, 125.

39. *Carroll v. FCC*, 258 F2d 440 (1958).

40. Bruce M. Owen, Jack H. Beebe, and Willard G. Manning, *Television Economics* (Lexington, Mass.: Heath, 1974), 145.

41. Michael Kinsley, *Outer Space and Inner Sanctums* (New York: John Wiley, 1976), esp. ch. 6-7.

42. FCC, *New Television Networks*, I 439.

43. Owen *et al.*, 62 ff.

44. FCC, *New Television Networks*, I 299-440.

45. See Posner for discussion of this "taxation by regulation."

46. Thomas Hazlett, "The Viewer is the Loser," *Reason*, July 1982.

47. 13 FCC 2d 420 (1968).

48. See Bob Kuttner, "Ma Bell's Broken Home," *New Republic*, March 17, 1982, for a summary.

49. E.g., *MCI v. FCC*, 561 F2d 365 (1977).

50. See John R. Meyer *et al.*, *Economics of Competition in the Telecommunications Industry* (Cambridge, Mass.: Oelgeschlager, Gunn, & Hain, 1980), esp. 119-22, 184-92, for an attempt to measure these factors.

51. Henry Geller, "Telecommunications," *Regulation*, Feb. 1982.

52. Barnouw, 139-40.

53. *KFKB Broadcasting Ass'n. v. FRC*, 47 F2d 670 (1931); and *Trinity Methodist Church, South v. FRC*, 62 F2d 850 (1932).

54. 319 U. S. at 215 (1943).

55. See Bernard Schwartz, "Comparative Television and the Chancellor's Foot," *Georgetown Law Journal 47* (1959), esp. 689 ff.

56. T. I. Emerson, *Toward a General Theory of the First Amendment* (New York: Random House, 1967), ch. 1.

57. E.g., *In the Matter of Editorializing by Broadcast Licensees*, 13 FCC 1246 (1949); and *Applicability of the Fairness Doctrine in the Handling of Controversial Issues of Public Importance* (the "Fairness Primer"), 29 Fed. Reg. 10416 (1964).

58. *Red Lion Broadcasting Co. v. FCC*, 395 U.S. at 390 (1969).

59. *Ascertainment of Community Problems by Broadcast Licensees, Primer*, 41 Fed. Reg. 1381 (1976).

60. Quoted from Great Lakes case in FCC, *Public Service Responsibilities of Broadcast Licensees* (the "Blue Book"), 1946, 12.

61. See documents cited in note 57.

62. 47 C.F.R. secs. 73.1920 and 73.1930.

63. Benno C. Schmidt, Jr., *Freedom of the Press vs. Public Access* (New York: Praeger, 1976), 170.

64. E.g., *CBS v. Democratic National Committee*, 412 U. S. 94 (1973); and FCC, *Report and Statement of Policy*, 44 FCC 2203 (1960).

65. Terms were changed in 1981 to five years in television and seven in radio.

66. *FCC et al. v. WNCN Listeners Guild*, 450 U. S. 582 (1981); "Reversing the D. C. Circuit at the FCC," *Regulation* May/June 1981; Winston Martin, "Decline and Fall of Broadcasters' Rights," *Reason*, Feb. 1979.

67. 359 F2d at 1009 (1966).

68. Schmidt, 190.

69. Bill Monroe, "Unchain the Electronic Media," *Reason*, Feb. 1979.

70. FCC, "Fairness Primer."

71. Schmidt, 170-71. Steven Simmons, *The Fairness Doctrine and the Media* (Berkeley: University of California Press, 1978), 80ff and Ch. 7.

72. *Comment in Wilderness Society Proc.*, 31 FCC 2d at 738 (1971), quoted in Simmons, 204.

73. An amendment in 1972 created an affirmative obligation to provide candidates with time; we discuss it in the next chapter. It has rarely been enforced.

74. In *Rep. Patsy Mink*, 59 FCC 2d 987 (1976).

75. Schmidt, 45-46.

76. Simmons, 210-214.

77. Ibid.

78. Ralph Jennings, quoted in Marvin Barrett, "Survey of Broadcast Journalism," *Columbia Journalism Rev.*, May/June 1979.

79. Simmons, 217.

80. The "Cullman doctrine" requires that free time be given if a sponsor cannot be found for the opposing viewpoint.

81. 390 U. S. at 393 (1969).

82. Fred Friendly, *The Good Guys, The Bad Guys, and the First Amendment* (New York: Random House, 1977), 39.

83. *Miami Herald Publishing Co. v. Tornillo*, 418 U. S. 241 (1974).

84. Jerome Barron, "Access to the Press—A New First Amendment Right," *Harvard Law Rev. 80* (1970), 1651.

85. Martin Mayer, *About Television* (New York: Harper & Row, 1972), 388.

86. Schmidt, 33.

87. *CBS v. FCC*, 453 U. S. 367 (1981).

88. Ibid., at 419.

89. FCC, *Annual Report for FY 1980*, 90.

90. U. S. Dept. of Commerce, Bureau of the Census, *Statistical Abstracts of the U. S.*, 1981, Table 971.

91. Schmidt, 38.

92. Anthony Smith, *Goodbye Gutenberg* (New York: Oxford University Press, 1980), 45-47, 62.

93. On the left, see Herbert Schiller, *The Mind Managers* (Boston: Beacon Press, 1973). Among conservatives, Kevin Phillips has been most vocal; see esp. his remarks in AEI's Roundtable, *Freedom of the Press* (Washington: American Enterprise Institute, 1976).

94. *U. S. News and World Report*, April 22, 1974, 34.

95. This research is reviewed in William C. Adams, "Network News Research in Perspective," in William Adams and Fay Schreibman (eds.) *Television Network News* (Washington: George Washington University, 1978).

96. Michael Robinson, "American Political Legitimacy in an Era of Electronic Journalism," in Douglass Cater and Richard Adler (eds.) *Television as a Social Force* (New York: Praeger, 1975), 115-16.

97. Simmons, 191.

98. Robinson, 116.

99. Simmons, 210-11.